Explorers of the Atom

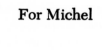

For Michel

Do not think for a moment that you know the real atom. The atom is an idea, a theory, a hypothesis. It is whatever you need to account for the facts of experience. [As our ideas about the atom have changed in the past, so will they continue to change in the future.] An idea in science, remember, lasts only as long as it is useful.

—ALFRED ROMER

Other Books by Roy A. Gallant

Explorers
of the Atom

Roy A. Gallant

Doubleday & Company, Inc.
Garden City, New York

The author wishes to thank the Graphic Arts Department of The American Museum of Natural History for permission to use the diagrams occurring throughout this book. I also wish to thank those colleagues of mine on the Museum staff, including the editors of *Nature and Science*, who were kind enough to read and comment on various chapters of the manuscript when it was being written. My special thanks to Laurence Pringle for permission to use "Travels of an Atom," which appears on pages 66 and 67.

Illustration Credits

Diagrams, unless otherwise credited, were prepared by the Graphic Arts Department of The American Museum of Natural History.
Bettmann Archive, p. 38
Hanford Atomic Products Operation, Richland, Washington, page 73
Theodore K. Himelstein, p. 31
Oak Ridge National Laboratory, page 75
Science Photo/Graphics, Ltd., pages 13, 14, 16, 18, and 20
Morris Warman, New York Medical College, page 46
Wide World Photos, page 59

Contents

8 *Contents*

Explorers of the Atom

Searching for the "World-Stuff"

How many times have you used the word "atom"? Many times, probably. And perhaps you have at least some idea of the way scientists picture an atom.

For instance, you probably know that an atom has a central lump of matter with one or more tiny bits of matter moving around it. And you may even know that there are a few more than one hundred different kinds of atoms.

If you do know those things, then you know more than anyone knew about atoms until less than a hundred years ago, even though men of ancient Greece "invented" the atom. The man usually given credit for first thinking of atoms is Democritus, who lived almost 2,500 years ago. Actually, it was Democritus' teacher, Leucippus, who had the idea first.

The Tiniest Particles

Democritus taught his students that all things in the universe—from stars to rocks to fingernails—were made of tiny particles, which he called *atoms*. He said that if you kept hammering a rock into smaller and smaller pieces, the smallest possible piece of rock-matter would be an atom. But it would be so small that you couldn't see it, feel it, or weigh it. He said that atoms are hard and solid ball-like objects that cannot be broken apart or chipped into smaller pieces. In fact, the word "atom" is an old Greek word—*a-tomos*, which means "not-divisible."

As we do today, Democritus imagined that there were many different kinds of atoms. Some, he said, are very light and free to dart about this way and that, and they can move far apart

DEMOCRITUS' ATOMS

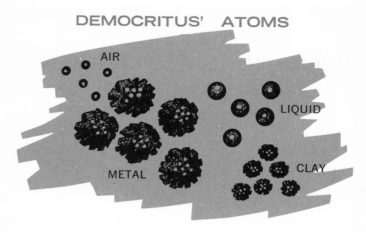

from each other. The air and other gases are made of such atoms, Democritus said.

But water had different kinds of atoms, thought Democritus, and they were arranged differently. He pictured the atoms of water and other liquids as larger and heavier than atoms of gases, because the atoms of liquids tend to stick together. And since anyone could see that liquids flow, their atoms must be slick and smooth. If they were not, they would not slip and slide over and around each other.

Atoms that make up copper, iron, rocks, and other heavy solid objects must be even larger and heavier than atoms of liquids, Democritus thought. And since it is hard to break apart such solid objects, their atoms must have very rough and jagged surfaces that cause the atoms to lock together tightly.

Democritus probably took that idea a step further by supposing that solids such as wood and soft clay are made of less jagged atoms whose surfaces do not lock together so strongly. He could explain the slow movement of "stiff" liquids, such as pitch oozing out of trees, in the same way.

For Democritus and a few others, the idea of a world made of atoms worked well enough. By thinking of all things as being made of atoms, they could explain the actions of liquids, solids, and gases, just as we can today.

But Democritus and the other Greek scholars of his time were thinkers, not experimenters. They were like mathemati-

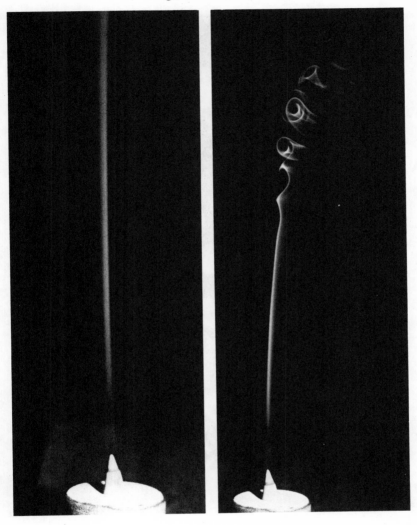

cians, not scientists. A mathematician can think out a problem, then solve it and prove the answer on the chalkboard. A scientist can also think out a problem and come up with an answer, but nearly always he must test the answer by doing experiments. The Greeks did not work that way. And since Democritus had only his thoughts to back up his theory, he had no way of proving that his ideas about matter were any "better" or more accurate than those of anyone else.

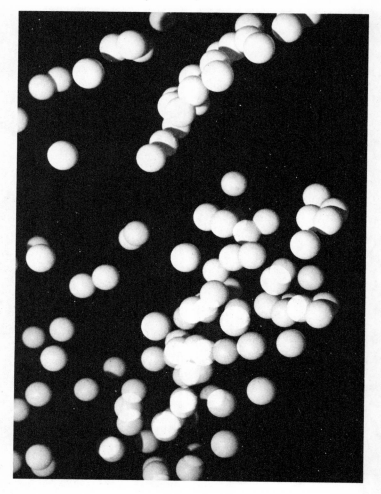

Earth, Air, Fire, and Water

As it turned out, the atom theory of Democritus did not catch on well at all. The idea about matter that did catch on was a very confusing one. According to that idea, there were four "elements"—earth, air, fire, and water. All things were made of one or more of these "elements." Wood, for instance, was made of earth, or so it seemed at first glance. If you heated wood, you could see that it also contained fire and air.

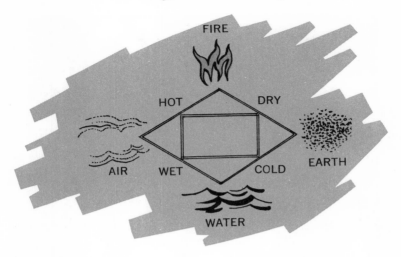

Perhaps the atoms of Democritus seemed too simple a way to explain something as big and important as the world. Anyway, the idea didn't catch on. The earth, air, fire, and water idea of elements did, and it stayed popular for about 2,000 years. Several men over those long centuries found atoms a little more convenient to explain the behavior of matter, so the idea did not die completely. But during all that time, no one came up with any strong objections to the four-element theory of what the "world-stuff" was made of.

Then in the 1600s and 1700s, discoveries came thick and fast. In the 1600s the English chemist Robert Boyle said that the four elements of the ancient Greeks couldn't begin to explain the many, many different kinds of matter around us. He felt that there must be many more "elements." When Boyle used the word "element," he meant what we mean when we use the word today—any substance that cannot be broken down into a simpler substance or built up from simpler substances.

Gold, silver, oxygen, and hydrogen, for example, are elements. On the other hand, carbon dioxide, the waste gas you breathe out, is made up of the two elements carbon and oxygen. Boyle had an idea that there must be many elements, but he could not guess how many. Today, we know of 105.

In the centuries when people believed that all substances were made up of earth, air, fire, and water, men called alchemists tried to find a way to change metals such as lead and mercury into gold. This drawing, made in the 1500s, shows an alchemist, working apparently in prayer for success in his experiments.

Elements, Atoms, and Molecules

John Dalton, an English chemist who lived a century after Boyle's time, took Boyle's idea of elements still further. He made the idea of atoms so convincing that few scientists of the time could doubt that atoms existed—even though they still couldn't see atoms, weigh them, or feel them.

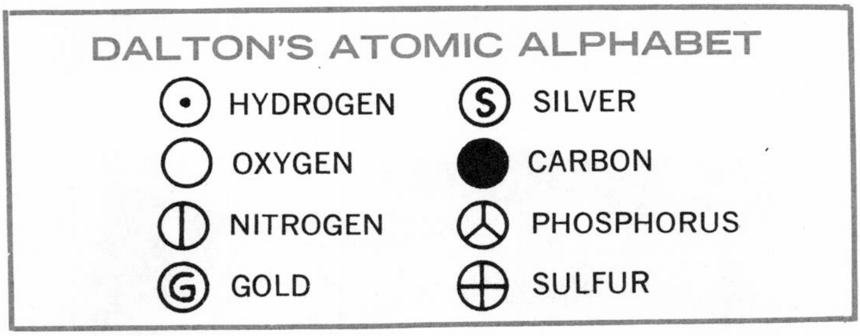

About the only "experimenting" Dalton was able to do with his atoms was on paper, not in a laboratory. He began with the idea that all atoms of any one element were exactly the same. The atoms of different elements were somehow different.

Dalton used a kind of picture writing to work out his ideas. As the diagram shows, he drew a black circle to stand for a carbon atom, an open circle to stand for an oxygen atom, and so on. Dalton thought that when matter changes—wood into fire, for example—atoms of one kind must join with or break away from atoms of another kind. He pictured a particle of water as being made up of one atom of the gas hydrogen joined to one atom of the gas oxygen, like this:

He called the pair a "complex atom."

Dalton's idea about atoms of hydrogen joining with atoms of oxygen and forming a new substance, water, was sound even though it was not correct. You have heard water called H_2O. That means that for each atom of oxygen in a particle of water, there are two atoms of hydrogen. If Dalton were around today, he would picture a particle of water like this:

H_2 O H_2O
HYDROGEN   OXYGEN   WATER

Today we imagine atoms as spheres of different sizes and mass. Different kinds of atoms join and form the many gases, liquids, and solids we find around us. The crystal solid shown here is a piece of calcite, a common mineral made of oxygen, calcium, and carbon atoms locked rigidly together in regularly repeating patterns.

Another scientist of Dalton's time gave the name *molecule* to Dalton's "complex atoms." He said that a molecule is the smallest possible amount of any substance that still acts exactly like larger amounts of the substance. For example, H_2O—one atom of oxygen joined to two atoms of hydrogen—is the smallest possible unit, or molecule, of water.

By 1844, when Dalton died, scientists all over the world had come to accept the idea of elements, atoms, and molecules. It was a very exciting idea, for it was bringing man the closest he had ever been to understanding what matter—including his own flesh and blood—was made of, and how it was put together. And eventually it was to lead men toward an understanding of life itself.

But at the time, perhaps even more exciting were these questions that began to be raised: How are atoms held together in a molecule? What force locks them together at one moment, then releases them the next moment? Would man ever be able to measure the size of atoms? Was it possible that atoms were not the solid, unbreakable objects that everyone pictured them to be?

Not only could such questions be asked, but answers could be found. For science had begun to move into the laboratory, where ideas could be tested and the results measured.

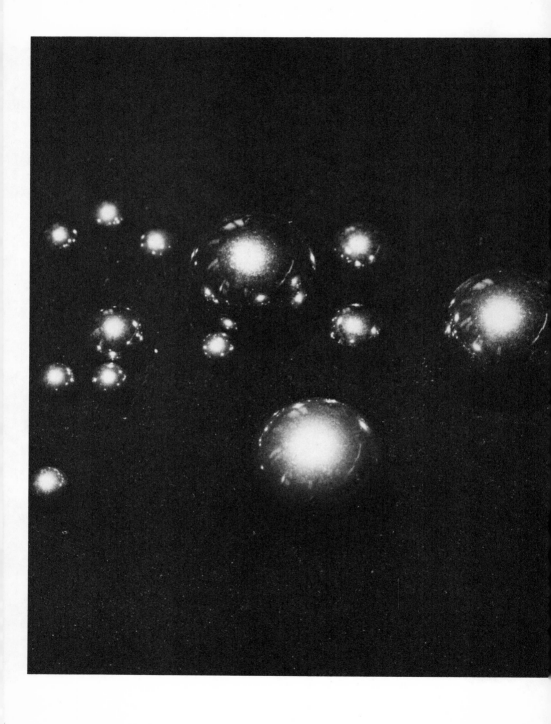

Man's First Look Inside the Atom

"I WAS BROUGHT UP to look at the atom as a nice hard fellow, red or gray in color, according to taste." That is how one scientist of the early 1900s, who was to make a major discovery about the atom, described those tiny bits of matter.

Up until almost 1900 the popular scientific view of atoms was that they were hard, solid spheres that could not be broken apart. But that view was soon to be changed. When it was, the atom began to open, and like a flower it revealed many hidden parts.

The Puzzling Green Glow

Our story begins in 1897 with the English physicist J. J. Thomson, who lived from 1856 to 1940. Thomson had been experimenting with electricity for about twenty years. One of the things he was trying to find out was what happens to electricity when it passes through a glass tube that has had most of the air pumped out. He could see that *something* was happening, but he could not explain it. As electricity passed through the vacuum tube, the tube glowed with a green light (Diagram 1, next page). Although scientists had known about this glow since Thomson was three years old, no one knew what it was.

Something was moving from one end of the tube to the other. But what? Just saying that it was "electricity" didn't explain anything. You could then ask, "Well, what is *electricity?*" The "something" seemed to be "rays" of some sort that traveled in straight lines. Thomson could see that they traveled in straight lines, because an object put inside the tube in the path of the rays would cast a sharp shadow (see Diagram 2). If the rays,

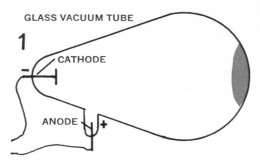

1 A glow of green light appears at the end of a glass vacuum tube when an electric current is passing through the tube between the cathode, or negative plate, and the anode, or positive plate.

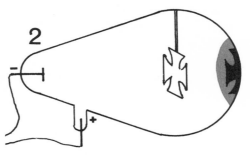

2 Thomson thought the glow must be caused by some kind of "rays" traveling straight out from the cathode, because an object placed in their path cast a sharp shadow on the end of the tube.

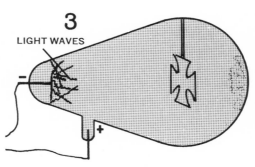

3 He knew that the "rays" could not be light waves from the heated cathode, because such waves would spread out in all directions, making most of the tube glow and producing only a blurry shadow behind the object in the tube.

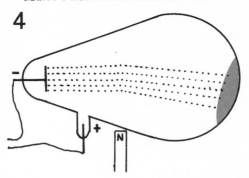

4 When the "rays" were pulled toward the end of a bar magnet held near the tube, Thomson concluded that they must be made up of a stream of particles, each particle carrying an electrical charge.

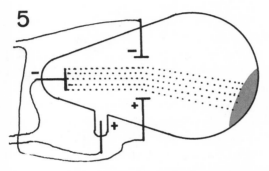

5 When Thomson passed the stream of particles between two electrically charged plates, the stream was bent away from the plate with the negative charge. This told him that each of the particles must carry a negative charge, since like charges were known to repel each other.

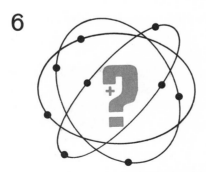

6 Thomson said that these negatively charged particles (electrons) are part of all atoms, but that some part of an atom must carry a positive electrical charge to balance the negative charges and hold the electrons.

or whatever they were, traveled in a helter-skelter way, then they would not cause an object to cast such a sharp shadow (see Diagram 3).

Thomson also used a magnet to try to find out more about the mysterious rays. When he held one end of a bar magnet near the tube, the rays were bent away from their straight-line path (see Diagram 4, where the "N," for "north" end of the magnet, is shown). While this did not tell Thomson what the rays *were*, it told him one thing the rays were *not*. They could not be rays of light, because light cannot be bent by a magnet.

Discovering Electrons

As Thomson experimented and puzzled over the problem, he was led to these two thoughts: (1) Perhaps whatever was passing through the tube was not rays at all, but instead was a stream of solid particles. (2) If so, then each of the particles must have an electrical charge, because their path can be bent by a magnet.

If he was right, then the mysterious stream would also be bent away from its straight-line path by electricity. When Thomson did an experiment like the one shown in Diagram 5, the same thing happened that had happened when he had used the magnet. When the mysterious stream passed between two metal plates with electricity in them, the stream was bent. This showed that the stream was made up of solid particles and that the particles were "charged." In other words, in some way the particles were being pushed away by the electricity, somewhat as the like poles of two bar magnets push each other apart. A stream of particles that were not charged would not be bent off course.

Thomson also managed to get at least some idea of the *mass,* or weight, of these charged particles. Meanwhile, other scientists had also managed to learn something about the masses of different kinds of atoms. For instance, they knew that hydrogen atoms were the lightest (least massive) of all known atoms, that oxygen atoms were heavier (more massive) than those of hydrogen, those of lead more massive than those of oxygen, and so on.

Thomson was able to say that his "corpuscles" of matter, as he called the particles, were much smaller than any known atom. He also said that these charged particles were part of all atoms. He even made a guess about how these particles might be arranged in an atom. He imagined each one to travel in a circular path, much as the planets circle the Sun. Today we call these particles *electrons*. The electricity (mysterious rays) Thomson had been experimenting with in the vacuum tube was a stream of electrons.

Thomson had broken the not-so-solid atom and had shown that it had tiny chips of particles of negative electricity. But what was the rest of the atom like (Diagram 6)?

An atom could not be made up *only* of electrons, Thomson reasoned. If it were, the electrons would scatter all over in an instant because they would repel, or push, each other just as the like poles of two bar magnets do. Also, the atoms making up a footstool, or a piece of chalk, or your skin, usually do not have an electric charge. Instead, they are electrically *neutral*, which means that some other part of the atom must have a positive charge that balances each electron's negative charge.

For each negative charge of each electron in an atom, there must be a balancing positive charge in some other part of the atom. So thought Thomson. Otherwise, an atom would carry a charge.

But in what part of the atom were these other charges to be found? Was there one positively charged particle to balance each electron? If so, were the positive particles scattered around inside the circular paths of the individual electrons? Or were they lumped together at the center of the atom? Thomson didn't know, but he guessed that they were scattered around. As it turned out, his guess was wrong. It was left to one of his brightest students to find the correct answer to that question.

Rutherford Finds a Nucleus

In the early 1900s, Ernest Rutherford, a New Zealander, was experimenting with an element called thorium. Like uranium, thorium has atoms that are big and not very stable—bits

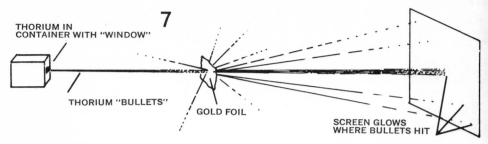

To explore the inside of atoms of gold, Rutherford aimed a stream of "bullets" from the element thorium at a thin piece of gold foil. A screen that glowed where a bullet struck showed that only a few of the bullets had changed course as they passed through the atoms of gold.

and pieces of them fly off from time to time. Rutherford knew that the tiny "bullets" that are shot out of a lump of thorium have a positive charge.

He used thorium for his exploration of the inside of atoms. As the diagram shows, Rutherford aimed a stream of thorium bullets (from the window opening of a container with thorium in it) at a thin sheet of gold foil. By finding out what the atoms of gold did to the tiny thorium bullets, he hoped to find out something about the part of the atom inside the "rim" of electrons. He could tell what happened to most of the thorium bullets by watching where they hit a special screen placed behind the gold foil.

Most of the thorium bullets followed a straight-line path and went right on through the barrier of gold atoms, just as if nothing were in the way at all. But some of the bullets seemed to "ricochet." If you did this experiment in a laboratory today, you would find that about one out of every 10,000 thorium bullets is knocked off course a bit when it passes through a sheet of gold foil. And once in an even greater while, one is knocked off course at nearly a right angle.

Those experiments helped show Rutherford two important things about atoms:

(1) Since most of the thorium bullets went right through the sheet of gold without changing direction at all, the atoms of gold must be mostly empty space. If they were solid, or mostly solid, even the thinnest sheet of gold would stop the bullets or change the path of most of them.

(2) Since some of the thorium bullets shot off at sharp angles as they passed through the gold foil, the central core of an atom of gold must be *very* much more massive than the electrons moving around the core. Otherwise, the thorium bullets would simply push the core to one side and plow straight on through.

With such information, Rutherford could now picture an atom. If the central core, called the *nucleus,* were made as large as a pea, then its nearest electrons would be about fifty yards away! Think about that for a minute, and you will understand how an atom can be "mostly empty space."

Protons, Electrons, and Neutrons

Since the time of Thomson and Rutherford we have found out a lot more about atoms, and we do not picture them in quite the same way that Rutherford did. For instance, experiments show us that the nucleus of the simplest atom we know about is made of one particle (called a *proton*) that has a positive charge. One electron forms a "shell" of electricity around that proton. At times the electron may be closer to the proton-nucleus than at others.

Other atoms, such as helium and lithium (see diagram on next page), have other particles in their nuclei. In addition to having protons, they have particles called *neutrons.* Neutrons do not have any electric charge, but they add mass to the atom. Atoms ordinarily have the same number of protons as electrons, so usually they do not carry an electric charge.

So electrons, protons, and neutrons are the three *basic* building blocks of all atoms except hydrogen. Although scientists today know that an atom has still more particles, the three basic ones are the most important.

The work of Thomson and Rutherford took physicists a long way toward understanding how an atom is made up of protons,

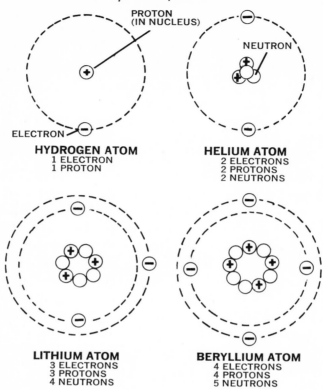

HYDROGEN ATOM
1 ELECTRON
1 PROTON

HELIUM ATOM
2 ELECTRONS
2 PROTONS
2 NEUTRONS

LITHIUM ATOM
3 ELECTRONS
3 PROTONS
4 NEUTRONS

BERYLLIUM ATOM
4 ELECTRONS
4 PROTONS
5 NEUTRONS

These diagrams show one atom of each of the four simplest elements. An atom normally has equal numbers of electrons and protons, so their negative (−) and positive (+) electrical charges are canceled out and the atom itself is without an electrical charge. Neutrons do not have a charge. They are packed tightly together with protons in the atom's nucleus, which makes up nearly all of an atom's mass ("weight"). Atoms with more electrons, protons, and neutrons than shown here make up the remainder of the 105 known elements.

electrons, and neutrons. It also opened the door to our under-standing of what happens to atoms when we say that they give off "radiation" or that they are "radioactive."

We are now in a position to look at some of the early ex-periments with radioactivity, which led scientists to discover some of the deadly dangers of the atom.

Case of the Mysterious Rays

THE TIME WAS 1896. The place Vienna, the capital of Austria. A newspaper story, which soon found its way onto front pages of other newspapers the world over, startled many readers.

According to the report, a professor in Würzburg, Germany, had found a way of making photographs of hidden things. Even the bones inside living animals could be photographed! Surgeons, especially, were interested in the new discovery. They now had an important new way of "seeing" a patient.

The name of the scientist was Wilhelm Konrad Roentgen. At the time Roentgen was doing his experiments, the British scientist J. J. Thomson was trying to find out what atoms were made of.

Thomson and Roentgen used the same kind of equipment, a cathode-ray tube. It was a glass globe or tube with most of the air pumped out. When electricity flowed from the cathode, or negative plate, to the anode, or positive plate, a stream of invisible rays caused that part of the tube struck by the rays to glow. Tubes made of English lime glass glowed with a green light. The tubes Roentgen used were made of German lead glass, which glowed with a blue light. Thomson had discovered, as we have seen, that the mysterious "rays" were not rays at all. Instead, they were tiny particles—high-speed electrons.

Discovering a New Kind of Ray

Roentgen's main interest was in the glow. No matter what part of a tube lit up with the bluish glow, rays of some kind were given off from the area that glowed. Roentgen discovered

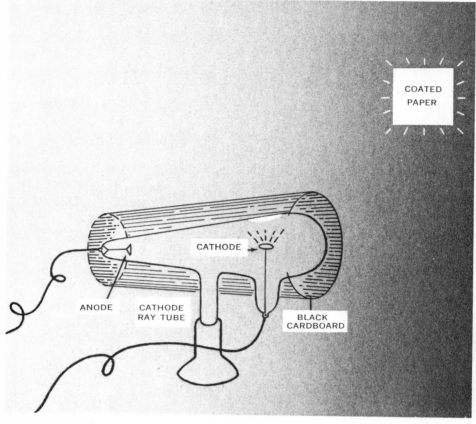

By accident, Roentgen discovered that some kind of invisible rays from a cathode-ray tube made the chemical coating on a sheet of paper glow with light. He tried to block the rays by covering the tube with black cardboard, but the rays passed through the cardboard and made the paper glow whenever electricity was flowing through the tube.

these rays almost by accident. To get a better look at the glowing tube, he turned out the room lights and fixed a thin, black cardboard shield around the tube to keep out any stray light. Off in another part of the room, there happened to be a sheet of paper coated with certain chemicals. Roentgen was surprised when he noticed that the paper sheet glowed each time the tube was turned on; but it did not glow when the tube was turned off.

He was even more surprised to find that the sheet of paper could be made to glow when it was moved into the next room. He suspected that the electric current passing through the tube made it give off rays of some sort. Roentgen's rays passed right through the glass and even through the thin wall of the room and so caused the sheet of paper in the next room to glow. He called them X rays.

During all of November and December of 1895, Roentgen worked day and night experimenting with the new rays. He found that they passed easily through thin and lightweight materials, but that they passed less easily through thick and heavy materials.

For example, in a darkened room he placed certain soft materials on a piece of photographic film and then aimed X rays at the film. He did the same thing with hard objects, such as metal keys and pieces of bone. In every case, the rays easily passed through the lightweight objects and exposed the film beneath them. But when he developed the film that was exposed with a key resting on it, he found an unexposed area the shape of the key where the key had blocked the X rays from reaching the film.

Roentgen placed a metal key on a photographic film in a dark room and aimed a beam of X rays at the film. The developed film showed an unexposed area the shape of the key where the X rays had been blocked from the film. (Also see photograph on page 33.)

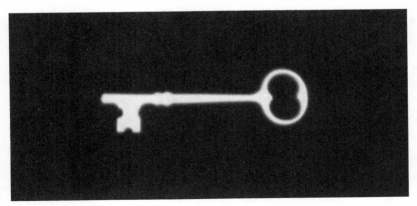

Announcing a Discovery

By Christmas 1895 Roentgen felt sure enough about his discovery of X rays to announce it to the world. When a scientist makes a discovery, he usually writes a complete report of his experiments, including any new ideas suggested by the measurements and other observations he made. He then has the report published in a scientific magazine, or he reads the report at an official meeting of some scientific organization. In that way, the scientist can feel pretty sure of being given credit for making the discovery.

There have been many times when a scientist made an important discovery but did not publish it soon enough. Meanwhile, another scientist in some other part of the world who was doing the same kind of experimenting published his report and got the credit for making the discovery, even though the other scientist had made it first. There have been many bitter arguments in science over who was the "first" to make a discovery.

Roentgen did not waste any time in making his discovery public. The Saturday after Christmas he had his report printed, then on New Year's Day he mailed copies to many of the leading scientists in Europe. Along with each copy, he sent X-ray photographs he had made. The copy that was sent to Vienna reached a newspaper editor. And the newspaper stories that followed made Roentgen's name and discovery known overnight.

The Scientist Who Reported Too Soon

The next act in this science drama took place in Paris only nineteen days after Roentgen had mailed off his report. One of the scientists in the audience listening to a report about Roentgen's X rays was a French physicist named Henri Becquerel. The more he thought about X rays, the more he began to wonder about them.

What interested him most was that the X rays came from that part of the glass cathode-ray tube that glowed. Such a glow is called *fluorescence*. Becquerel had known about other substances that were fluorescent, or glowed when exposed to

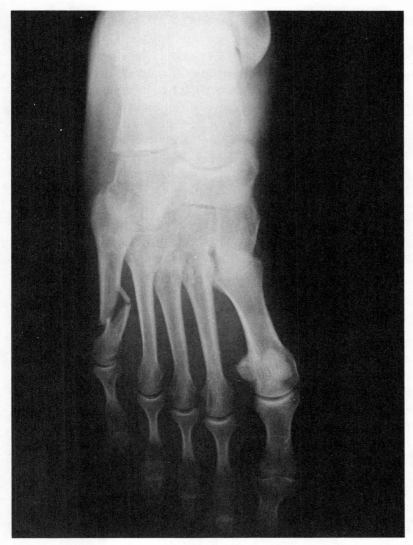

This X-ray photo of a human foot shows where a bone is broken. (X rays pass through flesh, but do not pass easily through bone.)

ordinary light. Could it be, Becquerel asked himself, that other substances that are fluorescent also give off X rays?

Here is an excellent example of how the work of one scientist can give someone else a new idea to investigate. Becquerel had asked an important question. His next task was to experiment and try to find an answer.

One day, he was experimenting with some crystals of a substance made of potassium, uranium, oxygen, and sulfur. For some time Becquerel had known that such crystals glowed when they were bathed in ultraviolet light. Ultraviolet light is that part of sunlight that causes sunburn. Its waves are too short to be detected by our eyes. To find out if the crystals gave off X rays when they were made to glow, here is what Becquerel did.

First he wrapped a sheet of unexposed photographic film in heavy black paper to protect it from the light. Next, he scattered several of the crystals on top of the paper and put the film package outside his window in the sunlight. Ultraviolet light from the Sun made the crystals fluoresce, or glow. After several hours, Becquerel brought the package inside and developed the film. In the exact places where the crystals had been resting above the film, he saw gray smudges!

Finally, it seemed, he was getting somewhere. He tried the same thing several times more. One time he would place a coin beneath one of the crystals. Another time he would place a piece of metal with holes punched through it beneath the crystals. Each time, he found a grayish patch in the shape of the object when the film was developed. On February 24 Becquerel read a report of his discovery to the French Academy of Sciences. In it, he said that he had discovered an X-ray-like ray that was produced by light.

Once a scientist makes an official announcement, such as Becquerel did, it can never be erased from his record. He must live with it always—right or wrong though his idea may be.

The experiments seemed foolproof. He had started with the idea, or *hypothesis*, that X rays are a regular part of fluorescence. Next, he experimented. His experiments then turned up exactly what he had predicted would happen. A beautiful piece of scientific thinking and experimenting—but completely wrong!

X Rays Without Light

Three days later, Becquerel discovered that he had been wrong. He was carrying out more of the same experiments,

Ultraviolet light (which you can't see) made certain minerals in this rock fluoresce, or glow, with a blue light, which appears white in this photo. When Becquerel detected X rays coming from a fluorescing rock, he mistakenly thought the X rays were caused by the ultraviolet light that was making the rock fluoresce.

but the weather turned cloudy and dark. Annoyed, Becquerel put one of the film packages he had carefully prepared into a drawer and tossed some of the crystals containing uranium on top of the wrapping. Then he closed the drawer. Surely the film would be safe from all light there, Becquerel thought. And the crystals could not possibly glow since ultraviolet light from the Sun could not get to them in the drawer. When it turned sunny again, he would then take the film package out into the light as before.

But Becquerel grew impatient. During those few cloudy days he decided to develop the film anyway. Imagine his surprise when he found patches on the film that were much sharper and darker than before.

Somehow, even without light, the crystals were giving off rays of their own. Becquerel repeated the experiment many more times. Each time, the same thing happened. No matter how long the crystals were kept in the pitch dark—hours, days, or weeks—they still gave off the mysterious rays.

Becquerel tested different kinds of fluorescent substances one after another. Those that were made of calcium or zinc, for instance, did not give off the rays. But every substance he tested that contained uranium did give off the rays. Uranium seemed to be the key substance. To make sure, Becquerel next used the film test on substances that contained uranium but that were not fluorescent. Each of these substances gave off rays that exposed the photographic film.

Becquerel now knew that neither visible light nor ultraviolet light were needed to make uranium give off rays. Somehow, the uranium was giving off rays all the time. Becquerel heated and melted his uranium substances, wondering if that would destroy the rays. It didn't. The melted substances, in liquid form, still gave off rays.

If uranium was, in fact, the substance that gave off the rays, then a pure lump of uranium might emit rays much more strongly than a substance that was only part uranium. At that time, in 1896, it was not easy to get pure uranium although more than fifty years earlier another French scientist had become the first to get pure uranium by separating it from the rocks that contained it. But a chemist Becquerel knew happened to be experimenting with pure uranium and provided Becquerel with a tiny amount. He placed a lump of the pure uranium metal on a piece of film wrapped in heavy black paper, as before. The results were the same! But this time the rays were the strongest ever. There could be no doubt now.

Uranium was a storehouse of energy and gave off rays, in the dark as well as in the light.

Becquerel, given an idea by Roentgen, an unknown scientist in another country, had made an important discovery about matter. It was a discovery that was to help other scientists come still closer to an understanding of what atoms are and how their energy can be put to work by man.

This photo shows Marie and Pierre Curie in their laboratory in Paris, where they discovered the elements polonium and radium. Marie Curie became the world's most famous woman scientist. She was the only person to win two Nobel science prizes—science's highest award.

New Atoms from Old

AT FIRST no one seemed to take much interest in the mysterious rays Becquerel had discovered coming from uranium. Roentgen's X rays had, for the time being at least, stolen the spotlight in science. With X rays one could make interesting photographs of the bones inside living organisms and of other "invisible" things. Becquerel's rays did not act in such a dramatic way.

It may be because Becquerel's rays seemed to be so uninteresting to other scientists that a young woman student at the Sorbonne decided to study them. That was in 1897, more than a year after Becquerel's discovery. She was a Polish girl born in 1867 and named Marya Sklodowska. In 1895 she had married a French physics professor, Pierre Curie and changed her name to Marie Curie. She was to become one of the world's most famous of scientists and to win not one but *two* Nobel prizes, science's highest award.

A Project for Her Doctor's Degree

By the year 1897 Marie Curie had already earned two university degrees, one in physics and one in mathematics, and was ready to start work on the highest degree of all, a doctor's degree. To earn one meant doing a very difficult and long research project. Becquerel's mysterious rays seemed a perfect topic for her research. No one else seemed interested in finding out what the rays were, so there would be a pretty good chance that another scientist would not try to solve the mystery before she could. It was a gamble, but it turned out to be a good one.

Becquerel had roughly measured the strength of uranium rays by observing by how much the rays clouded photographic film. Marie Curie wanted to measure their strength more exactly than that. One thing she knew was that the rays did something to the air they passed through, making it easier for electricity to flow through it. If she had an instrument sensitive enough to measure a small change in the flow of electricity through the air, she would have an excellent measure of the strength of the rays passing through. It so happened that her husband and his brother Jacques had made such an instrument (an *electrometer*). Now she could begin work.

She started by trying to find out, as Becquerel had, how a piece of pure uranium metal got the energy that it gave off in the form of rays. No matter what she did to the metal—exposing it to X rays, to ultraviolet light, or to heat—the strength of its rays remained the same. Perhaps other metals also gave off rays. She wondered. One by one, Marie Curie tried all the metals known to her. No luck, except in one case—the mineral ore pitchblende. But she had expected pitchblende to give off the rays. After all, it was the ore that contains uranium, so why shouldn't it? Pure uranium was obtained by separating the uranium from the rock part of pitchblende.

Discovering a "New" Element

What surprised her was that pitchblende gave off *stronger* rays than pure uranium did. How could that be? A gram of pitchblende should certainly give off weaker rays than a gram of pure uranium. That should be the case simply because the gram of pitchblende, with its rock and other impurities, has less uranium. A puzzled Marie Curie tested samples of pitchblende in every way she could think of. The answer was always the same—the pitchblende gave off the mysterious rays more strongly than pure uranium did.

Could it be that some other substance in the pitchblende was also giving off rays? Here was a new idea to test. For months Marie and her husband tested sample after sample of pitchblendes. They were trying to separate the pitchblende into each of its several different elements and to isolate the one emitting the strong rays. Gradually, they managed to do it. In July 1898 they ended up with a gray smudge at one end of a test tube. It was not uranium, and it was not any other element known at the time. But it gave off the rays, which Marie later called *radioactivity*. The smudge was in fact a newly discovered element. Marie decided to call it *polonium*, after her country of birth.

During their work, the Curies had detected still another radioactive substance in the pitchblende. In December of 1898 they managed to distil a tiny sample of mineral salt containing this substance. It was another "new" element, one that was about two million times more radioactive than uranium! They later named it *radium*. If they had known the trouble it would be to separate enough radium from pitchblende ore to find out how heavy radium is compared with the other elements, they would have been very disheartened. It turned out that they had to process eight tons of pitchblende to get only a few grams of radium.

Self-changing Atoms

By the early 1900s several scientists—in France, Germany, and Canada—had become fascinated by the newly discovered energy called "radioactivity," and, like the Curies, they all began experimenting with it. What kept the many experiments going, according to the physicist Alfred Romer, was publication: "Although a scientist may publish to get the credit of a discovery, the profit to all the rest of the world lies in the information he makes available, information from which anyone may pick up an idea about something even newer to try."

New ideas were now being published thick and fast. In England, Sir William Crookes was experimenting with uranium and discovered a puzzling new substance. It was so much like uranium that he called it *uranium X*. In Canada, Ernest Rutherford and Frederick Soddy were experimenting with the radioactive element thorium. During their investigations, they, too, came across what seemed to be a new substance. It was so like thorium that they called it *thorium X*.

Was it possible that the atoms of thorium and uranium had

the ability to change from thorium into thorium X and from uranium into uranium X? Rutherford and Soddy felt that is exactly what happened. At first the Curies had said no. The radium they had been experimenting with, and which had been in storage for months, showed no signs of changing. It was puzzling.

Rutherford and Soddy stuck with their idea that certain kinds of atoms, all by themselves, could change the way they were put together and in the process release energy as radiation. At the very moment such a change took place, the atom would give off a burst of energy as radiation. At least, that is how they pictured things.

Tracking Down the "Rays"

What, exactly, was this radiation? Was it bits of matter, like cathode rays? Waves of energy, like light waves? Or what? Rutherford found out by passing the rays from different radioactive elements through a magnetic field (see diagram). Some of the rays were bent sharply in one direction. Those rays turned out to be fast-moving electrons, the lightweight particles that J. J. Thomson had said make up part of all atoms. Rutherford called them *beta* particles.

But there were two more kinds of rays to be explained. One kind went straight through the magnetic field without being bent. Rutherford called them *gamma* rays. Since they were not pulled off course by the magnets, they were probably not particles of matter but waves of energy instead. They seemed to be extremely powerful—even more powerful than X rays— and they came from only a few radioactive elements.

The third kind of rays were bent slightly by the magnetic field, which meant that they must be moving particles, not waves of energy. Furthermore, they were bent in a direction opposite to the beta particles. Because beta particles have a negative electric charge, the opposite bending of these particles meant that they must carry a positive electric charge. Rutherford called them *alpha* particles. Since the path of the alpha particles was bent only a little, it seemed that they were much more massive (heavier) than beta particles. It turned out that

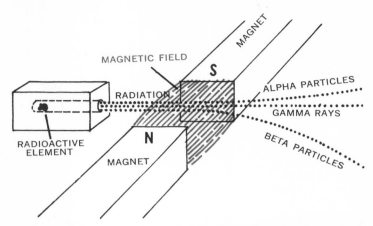

This diagram shows how the rays given off by radioactive elements act when they are passed through a magnetic field—the space through which the opposite poles of two magnets "pull" on each other. Beta particles (electrons) are bent sharply in one direction. Alpha particles (clumps of two protons and two neutrons each) are bent only a little, in the opposite direction from beta particles. Gamma rays (waves of energy, like light waves) are not bent at all.

they were four times as massive as protons, the particles that had earlier been found in the nucleus, or center part, of atoms.

Could that mean that each alpha particle was a clump of four protons? If so, then an alpha particle should have a positive electric charge four times as strong as a single proton. But it didn't; each alpha particle had a positive charge equal to that of only two protons. Why, if there were four particles?

Balancing the Alpha Particle's "Charge" Account

One possible answer was that two of the four protons of an alpha particle each had one electron sticking to it. The negative charges of the two electrons would cancel out the positive charges of the two protons. That would leave the alpha particle with the two positive charges of the other two protons (see diagram on next page). Electrons are so tiny that they would add hardly any mass at all to the alpha particle.

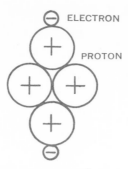

ELECTRON

PROTON

For thirty years scientists thought an alpha particle was made up of four protons and two electrons, with the electrons' negative charges canceling the positive charges of two protons.

This was an easy explanation, and it seemed to explain what made alpha particles behave the way they do. But Rutherford was not very happy with the idea. To him, a particle of radiation made up of six pieces of matter seemed too clumsy. About thirty years passed, though, before someone came up with a better explanation of the alpha particle.

In 1932 the English physicist James Chadwick did an experiment suggested by Rutherford and discovered another kind of particle that makes up part of atoms. This particle turned out to have just about the same mass as a proton, but it did not have any electric charge at all. Because it was electrically neutral, it was called a *neutron*.

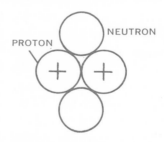

NEUTRON

PROTON

When Chadwick discovered the neutron—a particle with the mass of a proton but without an electrical charge, Heisenberg realized that an alpha particle is made up of two protons and two neutrons.

KIND OF ATOM	WEIGHT	NUMBER OF NEUTRONS	NUMBER OF PROTONS	PARTICLE LOST
URANIUM ↓	238	146	92	α
THORIUM ┊↓	234	144	90	β
LEAD	206	124	82	STABLE

Immediately, the German physicist Werner Karl Heisenberg saw the importance of Chadwick's discovery. An alpha particle, he said, did not have two electrons stuck to two of its four protons. Instead, each alpha particle was made of two protons and two neutrons. That explained why an alpha particle had the positive charge of two protons and the mass of four protons (see diagram).

Atoms and Radioactivity

With the discovery of radiation and neutrons, man's view of the atom had changed dramatically since the 1700s. The atom could now be pictured as a cluster of matter made up of protons and neutrons packed together in a central core called the nucleus, and electrons moving around the nucleus. But the atom was more complex than that.

The atoms of certain elements sometimes throw off one or more pieces of themselves, and this changes them into atoms of a different element (see box). Such elements are called *radioactive elements,* and the particles and the waves of energy that shoot out of them are kinds of radiation. Some of this radiation is weak; the beta particles (electrons) can be stopped by a thin sheet of metal. Gamma rays, on the other hand, are very powerful, energetic, and dangerous.

None of the scientists trying to solve the puzzle of radioactivity around the early 1900s realized the dangers they were exposing themselves to; nor did they realize that decades later science would find ways of using radiation to benefit man. Marie Curie died in July 1934 at the age of sixty-seven, of leukemia, which is cancer of the blood. Undoubtedly, the disease was caused by her exposure to high-energy radiation during her years of scientific work. Since that time we have learned not only about many of the dangers of radiation, but also how to use radioactivity in ways that help us arrest certain cancers.

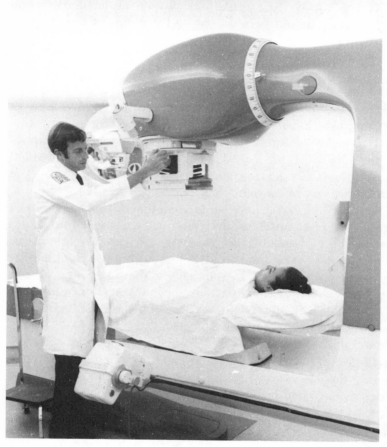

Man-made radioactive substances such as cobalt^{60} are useful in medicine. This machine at the New York Medical College in New York City contains a bit of cobalt^{60} and directs the radiation from its decaying atoms at a tumor deep in the patient's body. The radiation kills cells that make up cancerous growth.

Radioisotopes and How We Use Them

UNTIL ABOUT 1913 most scientists had thought that each atom of any particular element was exactly like every other atom of that element. According to this idea, each atom of gold was just the same as every other atom of gold; each atom of uranium was just the same as every other atom of uranium, and so on.

As early as 1886, however, at least one scientist had questioned the idea of sameness, but he could not prove it wrong. In 1912 two other scientists, J. J. Thomson and F. W. Aston, also began questioning the idea.

They were experimenting with atoms of the gas neon. First they stripped away electrons from the neon atoms. That left each atom with a positive electrical charge, so that the paths of the atoms would be bent as the atoms passed through a magnetic field. Next, they sent a stream of these charged neon atoms through a magnetic field. The stream of atoms curved and struck a sheet of unexposed photographic film (see diagram on next page).

If all of the atoms in the stream had been exactly alike, all would have followed the same curved path, and when the film was developed, there would have been only one gray patch where the atoms had struck the film. But there were *two* gray patches, one nearer the magnets than the other. That meant that some of the neon atoms were attracted by the magnetic field more than others were.

Thomson and Aston's experiment suggested that there were at least two different kinds of neon atoms, one kind more massive, or weightier, than the other. But the two scientists left the idea right there, failing to explore it more deeply. It was left to another scientist to see its importance.

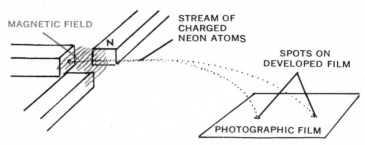

Thomson and Aston sent a stream of charged neon atoms through a magnetic field that pulled the atoms downward to a photographic film. The developed film had two gray spots instead of one, showing that some of the atoms were lighter than the others, because they were pulled downward in a sharper curve.

Weighing Atoms

The method Thomson and Aston used gave scientists a way of weighing atoms. Lighter charged atoms curved more sharply than heavier atoms as they passed through a magnetic field. By comparing the amount of curving, an experimenter could say that one kind of atom was two, or fifteen, or twenty-five, or whatever times as heavy as another kind.

At that time, the hydrogen atom was used as a standard unit of atomic weight and was given a weight of 1. Using that weight scale, Thomson and Aston's light neon atoms could be given a weight of 20 units and the heavy neon atoms a weight of 22 units. That is to say, the lighter neon atoms were twenty times as massive as a hydrogen atom, while the heavier neon atoms were twenty-two times as massive as a hydrogen atom.

At about the same time that Thomson and Aston were experimenting with neon, the English chemist Frederick Soddy was experimenting with the radioactive element thorium. His experiments showed that thorium was made up of at least two different kinds of atoms—one kind with an atomic weight of 232 and the other with a weight of 228. Even though they were different in weight, the two kinds of atoms were so alike in other ways that Soddy felt they should be thought of as the same element. Here was the important idea overlooked by Thomson and Aston. Soddy called such substances, whose atoms are alike in every way but weight, *isotopes*.

Let's take a look at isotopes of the element carbon. Every whole atom in the universe that has six protons and six electrons is a carbon atom. That makes each of those atoms enough like every other one that all of them can be called "carbon." But there is one kind of carbon atom that weighs 10 units, another that weighs 11 units, another that weighs 12 units, and so on, as shown here. Notice that carbon^{12}, for example, weighs more than carbon^{10} because carbon^{12} has two more neutrons. The important difference between isotopes of the same element, then, is the number of neutrons they have.

Isotope of Carbon	Number of Protons	Number of Electrons	Number of Neutrons
Carbon^{10}	6	6	4
Carbon^{11}	6	6	5
Carbon^{12}	6	6	6
Carbon^{13}	6	6	7
Carbon^{14}	6	6	8

With the idea of isotopes, it became easy to see why one cubic inch of lead, say, from one part of the world might weigh a little bit more or a little bit less than one cubic inch of lead from another part of the world. Each sample might contain different mixtures of lead isotopes. If each contained exactly the same mixture, then each sample would weigh exactly the same as every other sample.

Man-made Isotopes

In 1934 Marie Curie's daughter Irène and Irène's husband, Frédéric Joliot-Curie made an important new break-through. They were the first to make a man-made isotope. They did it by shooting alpha particles at aluminum. Their alpha-particle "gun" was a lump of the radioactive element polonium, whose atoms give off alpha particles.

Ordinary aluminum has an atomic weight of 27, because each atom has thirteen protons and fourteen neutrons. As the alpha particles bombarded the aluminum, the Joliot-Curies saw that the aluminum gave off neutrons and electrons. When they stopped the bombardment, they were surprised to find that the aluminum kept giving off electrons. They had made the aluminum radioactive. Each aluminum atom struck by an alpha particle gained two protons and one neutron. Because the atoms gained protons, they became the atoms of a different element. They were changed from aluminum to phosphorous. But this was not ordinary phosphorous (phosphorous^{31}), which has fifteen protons and sixteen neutrons. It turned out to be an artificial isotope of phosphorous—a radioactive isotope with a *half-life* of 3¼ minutes (see diagram).

The half-life of a radioactive isotope is the time it takes for half the atoms in a sample of the substance to decay, or change into atoms of another isotope. For example, in 4½ billion years, half the uranium^{238} atoms in a piece of rock change into atoms of lead^{206}. After 4½ billion years more, half the remaining uranium^{238} atoms will have decayed, and so on. Other radioactive isotopes have half-lives measured in thousands of years, in days, in hours, or in fractions of a second.

Since that important discovery made by the Joliot-Curies in 1934, about a thousand artificial isotopes have been made by scientists. All are radioactive, and their atoms decay at such rapid rates that none of these artificial isotopes is found in nature. The half-lives of some are measured in minutes; of others, in fractions of a second.

Measuring the Past with Geologic Clocks

Some radioactive isotopes have very long half-lives, so they can be used as "geologic clocks." They can tell us the age of ancient rocks and of fossils, and they can tell us something about the age of Earth itself. As you found earlier, the atoms of uranium[238], for example, gradually break down and change into atoms of other elements, ending up as lead[206].

It takes 4,510 million years for half the uranium[238] in a newly formed rock to change into lead[206]. By comparing the amount of uranium[238] with the amount of lead[206] in a rock, we can say that the rock is so many millions of years old. So far, the oldest Earth rocks dated with uranium[238] are about 3,800 million years old.

Since radioactive potassium[40] is found in several common minerals, such as feldspar and mica, it is becoming a popular way of dating rocks. Half of the potassium[40] in a newly formed rock breaks down into argon[40] in 1,350 million years.

The half-lives of some isotopes are too long, though, to tell us the age of fossils that were part of living plants and animals within the past 50,000 years or so. For this purpose, another radioactive isotope is used—carbon[14].

This isotope has a half-life of only 5,760 years, so you might think that all the carbon[14] on Earth would have disappeared long ago. But new carbon[14] is being made all the time as cosmic rays—powerful bundles of energy from outer space—change nitrogen[14] in the upper part of Earth's atmosphere into carbon[14]. The "new" carbon[14] atoms replace those that have decayed, and the tiny fraction of the carbon in the atmosphere that is carbon[14] seems to stay the same.

Living plants take in the isotope carbon[14] from the atmosphere, and living animals get carbon[14] from the plants or from the plant-eating animals they in turn eat. So carbon[14] makes up the same fraction of the carbon in living plants and animals as it does of the carbon in the atmosphere. When a plant or animal dies, it stops taking in carbon. The fraction of the carbon in its remains that is carbon[14] gradually gets smaller as the atoms of carbon[14] decay and change into atoms of nitrogen[14].

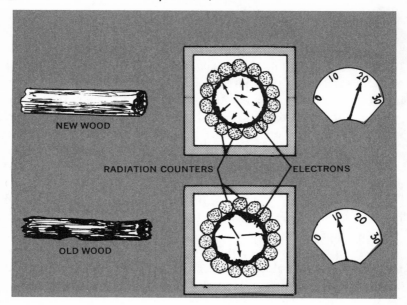

An ancient piece of wood can be dated by comparing the carbon[14] radiation coming from it with the radiation coming from a piece of wood the same size and weight recently cut from a living tree. If radiation counters show that the new wood gives off twenty signals per minute, and the old wood only ten signals per minute, the old wood must contain only half as many carbon[14] atoms as the new wood. Since half the carbon[14] atoms in a piece of wood decay within 5,760 years after the wood is no longer part of a living tree, then the old wood must be about 5,700 years old. (If the old wood gave off only five signals per minute, how old would it be?)

You can detect how much carbon[14] there is in a piece of wood, say, by holding a radiation counter near it. If the piece of wood is about the size of a pencil and comes from a tree that was cut down a few weeks ago, the radiation counter will record about twenty "signals" a minute—each signal announcing the change of a carbon[14] atom into a nitrogen[14] atom.

Now suppose that you find a piece of charred wood in the remains of a fireplace used thousands of years ago. You cut off a piece that weighs the same as the piece of new wood whose carbon[14] count you already know. Next, you hold the radiation counter near the piece of old wood.

Suppose you count only ten signals a minute coming from the old wood. That would tell you that the old wood had only half as many carbon^{14} atoms as the new wood. It would also mean that the old wood had lost half the number of carbon^{14} atoms it had when it was part of a living tree. Since it takes 5,760 years for half the carbon^{14} atoms in a dead animal or plant to decay, you can say that the piece of charred wood was cut from a tree about 5,700 years ago.

Some Other Ways We Use Radioactive Isotopes

Over the years we have found many uses for radioactive isotopes. Their use in medicine has become particularly important. The high-energy radiation some isotopes give off can be directed at a tiny spot in the body to kill cancer cells.

Less powerful isotopes can be injected into a person's blood stream or be included in something we eat or drink. By tracing the signals given off by the isotope as it becomes trapped or absorbed by some tissue or organ along the way, doctors can learn about a person's blood circulation or about how well his body uses what he eats.

Such "tracer" isotopes, as they are called, can also help us find out how certain nutrients are used by plants. And the isotopes can help us study certain behaviors of insects and other animals. When scientists of the early 1900s did their work with radioactivity and isotopes, they probably never imagined what an important new tool they were putting into the hands of scientists of the next generation.

Biologists use the radioactive isotope strontium^{90} to help them count populations of small animals and trace their travels. This diagram shows how strontium^{90} can be used to count mosquito populations.

1 Strontium^{90} fed to mosquito larvae in laboratory tanks makes them grow into radioactive adults (colored dots) that can be detected by a radiation counter.

2 One hundred radioactive mosquitoes are released in an area where biologists want to count the mosquito population.

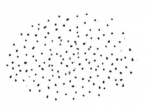

3 A sample of 100 mosquitoes captured in the area included only one-tenth of the radioactive insects, so there must be about 10 times 100, or 1,000, mosquitoes in the area.

Energy from the Atom

FOR ABOUT 2,000 years certain "scientists," known as alchemists had dreamed of changing base metals, such as lead and mercury, into gold. Surely, they had thought, it could be done if only they could hit on the "right mixture" of things. Unhappily, the "right mixture" was never found, at least not by the alchemists.

But scientists of the 1900s did find a way of making gold, and real gold. In 1941 an American physicist using an atom smasher made a few million atoms of gold out of atoms of mercury—a two-thousand-year-old dream come true. Five years earlier another physicist had managed to change platinum to gold. The physicists' success came not from discovering some "secret mixture" as imagined by the alchemists, but from rearranging the very building blocks that make up atoms —protons, neutrons, and electrons.

Discovering how to make gold was not very important. It cost far more to make the few million atoms of gold than the gold was worth. So it's still cheaper to dig it out of the ground than to make it in a laboratory. What was important was that scientists of the early 1900s were learning to reshape atoms. By so doing, they were well on their way to unleashing a rich source of energy from the nucleus of the atom, energy that could be used to move mountains.

Splitting the Atom

The scientist who first began to experiment along these lines was the Italian physicist Enrico Fermi (1901–54). From the work of others, Fermi knew about man-made isotopes. For example, if a neutron "bullet" strikes the core of an atom of oxygen[18], the atom changes into an atom of oxygen[19]. It then gives off a beta particle (an electron) and in the process changes into an atom of a different and heavier element, fluorine[19].

The heaviest element known in 1934, when Fermi was doing his experiments, was uranium. As you saw earlier, an atom of ordinary uranium has 92 protons and 146 neutrons, so its atomic weight is 238. Fermi wondered if he could change uranium atoms into atoms of an even heavier element by shooting neutrons into their nuclei. Although he succeeded in changing uranium into different substances, he could not identify those new substances.

Four years later, the German scientist Otto Hahn, who had joined in Fermi's investigations of uranium, came up with the explanation that Fermi had missed. But it seemed so wild that Hahn would not talk about it in public.

One substance produced by bombarding uranium with neutrons seemed to be an isotope of the element barium. But how could that be? Barium is much lighter than uranium. To make a barium atom from a uranium atom would mean that a uranium atom would have to be broken more or less in half. Surely, that was impossible. Or was it?

Here was the germ of an idea that was to give man a source of energy rich beyond his fondest dreams—and also a weapon of war fearsome beyond his imagination, the nuclear bomb.

It was the Austrian physicist Lise Meitner, a co-worker of Hahn's, who saw the importance of Hahn's and Fermi's work and did something about it. When uranium atoms are split, she reasoned, they break down and become atoms of certain other elements, among which are barium and krypton. But the combined mass of one barium and one krypton atom is less than the mass of one uranium atom. What happens to the "lost" mass? Meitner saw the answer: The missing mass is not lost at all. It is changed into energy!

Making Atoms Split Other Atoms

The time was World War II, when bombs were important. Otto Hahn had managed to bring about the *fission*, or splitting, of uranium atoms into atoms of barium and krypton. But this process was like lighting a match that went out right after it flamed up. How could a steady release of energy be kept going?

By this time, Fermi and his family had come to America to live, because he opposed the Fascist regime of his native Italy. Working at the University of Chicago, Fermi saw that the atomic match struck by Hahn *could* be kept going—if the right approach were used. It was Fermi who worked out that

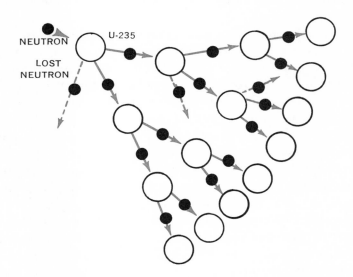

The fission chain reaction in an atomic pile begins when a neutron from a decaying atom of uranium235 (*top left*) strikes another U235 atom, making it split into two nearly equal parts. At the same time 3 neutrons and energy are given off. Some of the neutron bullets strike other U235 atoms, releasing more and more energy as more and more U235 atoms are split. The diagram here does not show all of the details of a chain reaction; its purpose is to show only that one split atom leads to the splitting of other atoms, which in turn split still other atoms, and so on.

approach by getting a pile of uranium atoms to split and keep on splitting, in what came to be called a *chain reaction*. During such a chain reaction great amounts of heat and other forms of energy are given off.

Fermi reasoned that when a uranium atom was split in two by a neutron bullet striking its nucleus, one or more neutrons were set free. If so, then why couldn't those neutrons become bullets to split other uranium atoms, and the neutrons from those atoms split still other atoms, and so on and on, in a chain of reactions?

The problem was to get a big enough pile of uranium so that there would be enough atoms to keep a chain reaction going. At the time—1942—there wasn't much uranium around. But with the help of the United States Government, which was secretly sponsoring the project, Fermi managed to get enough natural uranium to make a doorknob-shaped pile about twenty-four feet across.

Although uranium^{238} atoms can be split, they cannot keep a chain reaction going all by themselves. A special isotope of uranium is needed, uranium^{235}. Of one hundred atoms of uranium taken out of mines, ninety-nine are atoms of uranium^{238}; only one is a uranium^{235} atom. But the uranium pile made by Fermi was large enough so that there were enough uranium^{235} atoms to keep a chain reaction going, if only a weak one. Even though Fermi's atomic "furnace" didn't give off very much heat, it worked; that is, it kept going. It proved beyond a doubt that the atom could be harnessed as a powerful new source of energy—for peace as well as for war.

Meanwhile, Albert Einstein had written President Roosevelt that ". . . this new phenomenon would also lead to the construction of bombs, and it is conceivable—though much less certain—that extremely powerful bombs of a new type may be constructed." The nuclear bombing of the Japanese cities of Hiroshima and Nagasaki followed three years later, ending World War II. But those bombings also revealed to man the horrible destruction he could wreak on himself.

While we have made even more powerful bombs since then, we have also learned to use nuclear energy in peaceful ways. Many nuclear reactors (atomic piles) were built during the

This drawing shows the world's first atomic pile, built by Fermi and his co-workers in 1942 in a squash court under a stadium at the University of Chicago. The photo below shows how layers of graphite tile with uranium in the center were separated by layers of solid graphite tile to slow down the neutrons and keep them from escaping from the pile.

1950s and 1960s. Today they power submarines for war and turn generators that make electricity to light our homes.

Slowing Down a Chain Reaction

The main difference between a nuclear reactor and a fission bomb is the rate at which uranium atoms in the pile are split. In a bomb we just let the freed neutron bullets fly around as they will. The chain reaction then goes so fast that a huge amount of energy is released in an instant as an explosion.

In a nuclear reactor, we slow down the chain reaction by putting shields in the way of the freed neutron bullets. The shields may be blocks or rods of graphite or some other suitable material. The greater the number of shield rods that are inserted into a reactor, the greater the number of freed neutrons that are blocked; and so fewer atoms are split, resulting in still fewer neutron bullets. In this way the chain reaction is made to go slowly. Energy trickles out only as fast as we want it to, instead of exploding out. Pull out all the shield rods and the reactor would go wild, or "critical" as the scientist terms it.

In the 1950s a second kind of nuclear energy was unleashed by scientists. While nuclear reactors release energy by splitting apart very massive atoms, the new source of energy came from *fusing*, or joining, the nuclei of very light atoms. During the

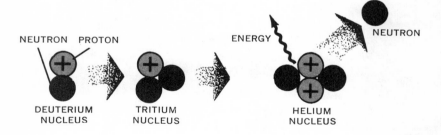

Intense heat makes the nuclei of heavy hydrogen atoms (deuterium and tritium) fuse, or combine, forming the nuclei of helium atoms and giving off a neutron. This fusion reaction releases more energy than the fission of a uranium atom does. The Sun's energy is produced by fusion reactions taking place deep within the Sun's core region.

fusion of two such atoms, much more energy is released than that released by splitting uranium atoms.

The Sun's great outpouring of energy comes from the fusion of hydrogen nuclei. So does the energy from H (for hydrogen)-bombs.

While uranium is the fuel for fission reactors, hydrogen is the fuel of fusion reactors. Actually, it is not ordinary hydrogen that is used in fusion reactors, but two isotopes of hydrogen. One is deuterium (which has one proton and one neutron in its nucleus, also called *heavy hydrogen*), and the other is tritium (which has one proton and two neutrons). When the nucleus of a tritium atom and the nucleus of a deuterium atom are fused, they form the nucleus of the slightly more massive atom of helium, which has two protons and two neutrons. At the same time, energy is given off and one neutron is set free.

Energy from Fusion

While slow neutrons can set off a fission reaction, much more energy is needed to set off a fusion reaction. A neutron bullet fired at a uranium nucleus is not repelled by the nucleus because the neutron does not have an electric charge. The positive charge of the protons in the nucleus does not act on the neutron at all. But think about the nuclei of two hydrogen atoms being fired at each other. Both are positively charged because each has a proton. That means that they repel each other. How, then, do we get them to join, or fuse? We must fire them at each other with enough energy to overcome their pushing each other apart.

The explosion of a fission bomb releases enough energy to set off a fusion reaction by ramming the hydrogen nuclei together. An H-bomb is nothing more than a fission bomb of heavy nuclei surrounded by a certain amount of lighter nuclei of deuterium and tritium. When the fission bomb goes off, it causes the hydrogen nuclei to fuse and release energy. It all happens so fast that we see both events as a single explosion.

As we have been able to slow down the energy release in fission reactions and control the energy to power machines, we are now trying to control the energy coming from fusion

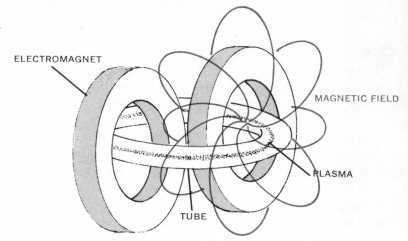

To produce a continuous, controlled fusion reaction, scientists must heat plasma—a gassy mixture of heavy-hydrogen nuclei—to very high temperatures while keeping the plasma squeezed together. One way this may be done is by sending electric sparks from strong electromagnetic coils around the tube through plasma held in the center of a tube by the magnetic field.

reactions. But so far we have not been able to. What it amounts to is making a miniature Sun in the laboratory.

Although we can start a safe fusion reaction in the laboratory, we cannot keep one going for more than about fifty thousandths of a second. It is like trying to start a match in a windy tunnel. You can strike it and get a flame, but each time it goes out. If we could keep a fusion reaction going for only one second, that would be long enough so that it would keep going. So far, the Russians have come the closest to making a fusion reactor that works. Some scientists think that within five or six years a successful fusion reactor may be operating in the United States.

There are two main advantages to fusion reactors over fission reactors. The fuel for a fusion reactor is heavy hydrogen (deuterium), which is found in the oceans. And there is plenty of it, enough to last us for millions of years. Like coal and oil, uranium for fission reactors is in much shorter supply. Another important advantage of fusion reactors is that they would not leave dangerous radioactive "ash." Fission reactors produce radioactive wastes, and disposing of these wastes is becoming a serious problem, the subject of of the next chapter.

Pollution from the Atom

IT HAS BEEN about thirty years now since the United States exploded the first atomic bomb over Japan. At that time, no one could foresee that man was about to become lethally involved with the most deadly of all pollutants—radiation. No one could know that in a short thirty years or so radiation pollution would be a planet-wide concern.

While scientists knew about many of the harmful effects of radiation, no one could know just how serious a threat to our health radioactive waste material from bomb tests and uranium processing and nuclear power plants would be over a period of many years. And we still do not know as much as we should about what happens to the radioactive wastes themselves once they foul the air we breathe and poison our lakes, rivers, and soil.

Radiation and People

Radioactive waste materials give off three types of radiation: (1) *Alpha particles*—each alpha particle is a cluster of two protons and two neutrons; these particles are relatively heavy and slow moving and are stopped by the skin, but we can breathe them in or swallow them when they are part of our food. (2) *Beta particles*—each beta particle is an electron; these particles are more energetic than alpha particles and can pass through flesh or several inches of wood. (3) *Gamma rays*—these are waves of energy and are the most damaging pollutant; gamma rays can pass through several inches of steel or several feet of concrete. What does all of this mean to you?

When a nuclear test bomb is exploded in the air, underground, or under water, it produces isotopes that give off all three types of radiation. And so do the millions upon millions of tons of waste material left over from the mining and processing of uranium for use in nuclear reactors and in bombs. These isotopes become part of the air we breathe, the food we eat, and the water we drink and use to water our crops. They then get stored up in our bones, in certain glands, and in other parts of our body, and there's no way we can give them up.

When a radioactive atom that we have breathed in from the air gives off a burst of damaging radiation, that radiation streaks through our soft body tissue, leaving a trail of damaged atoms in the cells it has passed through. It may damage normal atoms by knocking off electrons, leaving the atoms with an electrical charge. Such atoms may combine violently with other atoms or with molecules, killing or injuring the cell they are part of. After weeks, months, or years, cells injured in this way may begin to grow and reproduce in the wild manner that we call cancer. Radiation is a cause of cancer.

Cells damaged by radiation are affected in still other ways that we do not yet understand. If a damaged cell happens to be a human sperm cell or egg cell, which join and produce a new human being, the baby might be born deformed. Radiation also seems to make our bodies age faster, though the details of how this happens are not yet understood very well. It is one more of many ways in which we are ignorant about the dangerous effects of radiation.

When radioactive wastes are dumped into the air or into rivers or oceans or buried in the ground, they do not go away as garbage from food does. Our radioactive garbage stays around for tens, hundreds, thousands, or millions of years. Fallout from H-bomb explosions in the atmosphere is carried around the world by the winds and gradually settles to the ground. Meanwhile, we breathe it in. In 1963, realizing the dangers from excessive radioactive fallout, the U.S.S.R., Britain, and the U.S. agreed not to explode any more H-bombs in the atmosphere. France and China, however, have not joined the test ban.

Long-lived "Garbage"

Radiation from radioactive wastes dumped into rivers is taken in and stored by plants growing in the water. Even though there may be very few radioactive atoms in a glass of river water, the plants keep collecting and storing them. Eventually, a plant may contain several hundred times as many of the radioactive atoms in its tissues as there are in an equal volume of the surrounding water.

Small fish that eat the plants are next in line to take in the radiation pollutants. They, too, store up and concentrate the radiation in their bodies. Larger fish then eat the smaller fish. At this stage in the food chain the concentration of radioactive atoms in a bass, for instance, may be 1,000 times higher than it is in the surrounding water. Birds and man may then eat the bass. Each step along the way the new consumer is taking in the radiation stored all down the line. Collected, concentrated, and carried in this way by animals, radioactive pollutants dumped into a stream in Colorado or Washington may be carried by migrating birds or by man to another state or to another continent. It is a wild dream to suppose that radioactive wastes stay put. They don't.

As an example of what can happen to a radioisotope once it enters the food chain, let's consider the radioactive isotope zinc[65]. A government-operated nuclear reactor located near the Columbia River in Richland in the state of Washington for several years has been polluting the river by leaking zinc[65] into it. This radioactive element is taken up by microscopic plant organisms called plant plankton. The plant plankton store up the zinc[65], so while the radioactive element may not be very concentrated in the water itself, it becomes concentrated in the plant plankton. Now the plant plankton are gobbled up by tiny animals called animal plankton. The more plant plankton the animal plankton eat, the more the animal plankton store up zinc[65]. The animal plankton may store up ten times as much of the radioactive element in their body tissues as the plant plankton do. Small fish that eat the animal plankton may store up ten times as much of the zinc[65] as the animal plankton do; and larger fish that eat the smaller

1 Let's follow a carbon atom, called C. We'll pick up its trail 300 million years ago, when it was part of a giant insect like a dragonfly. The insect died and its body settled into the bottom of a swamp.

2 As the insect decayed, many of its atoms were arranged in new molecules and recycled into the air or water. But its body didn't decay completely. Many years passed and the insect's remains were buried deep under layers of dead leaves from the swamp plants. Eventually, pressure and heat caused by the weight of the material above produced changes in the bottom layers. The atoms of the insect's body and of the plant leaves were rearranged into new molecules, forming coal, a kind of rock. Coal is made up mostly of carbon atoms.

3 After many millions of years, most of the rock and soil above the layer of coal in which C rested was worn away. Roots from a grass plant found a watery crack in the coal and grew down into it, cracking it still wider. C joined with two oxygen atoms in the water, resulting in a molecule of carbon dioxide. The molecule was taken into the cells of the plant's roots and was carried from cell to cell up through the plant to its leaves.

TRAVELS OF AN ATOM   To grow and develop, living things need more than thirty different elements. Your body is made up of elements such as carbon, hydrogen, oxygen, nitrogen, sulfur, calcium, and iron. The same elements, in varying amounts, can be found in all living things—in a frog, an ant, a blade of grass.

Each element is made up of just one kind of atom. Atoms sometimes exist by themselves, but usually they combine with other atoms.

When a plant or animal dies, its atoms are not "lost." They are used again and again (recycled), joining with other atoms to become molecules in soil, air, water, and rock, and in other living things. Some of the atoms that make up your body may have been recycling since Earth began. Most of them came to your body in the food you've eaten, liquids you've drunk, or air you've breathed.

Most atoms were locked up in rocks for millions of years, others were miles deep in the ocean or miles high in the atmosphere. And some atoms were part of other living things. One of the atoms in your body may have once been part of a redwood tree, a mouse, or a dinosaur.

7 The rabbit's bones will eventually decay. C may remain in a bacterium for a time or it may join two oxygen atoms and escape into the air as carbon dioxide, or C may stay in the soil until a plant's root takes it up. Then if you happen to eat the plant, C will become part of you. Radioactive atoms released to the environment by man travel through the food chain just as the carbon atom (C) has in the example here.

6 The rabbit died several months later. Birds, insects, and other animals ate most of its flesh. Bacteria released chemicals that broke down big molecules into smaller ones. Then the small molecules passed through the cell walls of the bacteria. From these molecules, the bacteria got energy to live and grow. Soon only some hair and the rabbit's bones were left.

5 Protein molecules are complex, made of thousands of atoms. They are especially plentiful where growth is taking place. That is why C was near the growing tip of grass stem that was nipped off by a cottontail rabbit. Inside the rabbit, the molecules of grass were broken down and rearranged into molecules needed by the rabbit's body. C combined with atoms of oxygen and calcium to become a molecule of calcium carbonate in part of a rib bone.

4 In the green leaf cells, molecules of carbon dioxide and water are rearranged in a process called *photosynthesis*. Molecules of sugar are formed. The sugar may later be changed to molecules of starch, fat, or cellulose—all of which contain atoms of carbon, hydrogen, and oxygen. When these kinds of atoms are rearranged with nitrogen atoms, molecules of proteins are formed.

fish may have ten times as much zinc[65] as the smaller fish do. So a large fish may end up with $10 \times 10 \times 10 \times 10$, or 10,000 times the concentration of zinc[65] as the plant plankton, which form the first link in the food chain. Biologists are now trying to find out to what extent large fish near the upper end of the food chain may be damaged by radiation.

Oysters living at the mouth of the Columbia River also pick up the zinc[65], but the story doesn't end there. Columbia River water is used for irrigation. Plants taking up the water also concentrate the zinc[65] in their tissues, and so do milk cows and other grazing animals that eat the plants. Man, the end link in the food chain, also concentrates the harmful radioisotope in his body tissues, where it lingers for fourteen years. In this way, our bodies can be exposed to high concentrations of certain radioactive substances—all of which are harmful. We can be harmed by the substances even though the concentrations of them dumped into the environment are so low that they can be hardly detected at the time.

Here is still another example of how radioactive substances get around. Several years ago radiation counts made on a group of Eskimos living around Anaktuvuk Pass in northern Alaska showed that the Eskimos had absorbed more radioactive fallout from nuclear bomb tests than people living in more southern latitudes. It was puzzling, because the air over Arctic regions was thought to carry less fallout than air farther south. The puzzle was soon solved when someone thought of examining the Arctic food chain. It turned out that scrubby plants called lichen had been absorbing and concentrating radiation from the air. Lichens happen to be an important food for caribou. In turn, caribou happen to be an important food for the Eskimos. To this day the Anaktuvuk Pass Eskimos who eat caribou meat continue to take in man-made radiation concentrated by lichen growth. What long-term effects this radiation may have on them are unknown at the present.

In the normal course of eating and sleeping, our bodies take in and use the atoms of many different elements. While certain of those atoms become building blocks as our bodies grow new cells, certain other atoms are needed to convert the food we eat into energy. If there are radioactive atoms about, in our food and in the air around us, our bodies take them in right along with the atoms that are not radioactive.

Carbon is one such element, and our bodies cannot tell ordinary carbon (carbon^{12}) from radioactive carbon^{14}. In the small amount our bodies have become used to, radioactive carbon^{14} seems to be harmless and it is stored by nearly all of our body tissues. This isotope has a half-life of about 5,700 years. Strontium^{90}, a radioactive isotope from H-bomb fallout, is another radioactive element we take into our bodies. Our body tissues mistake strontium^{90} for calcium and use it as a bone-building material. Since growing children are actively building bone tissue, they stand more of a chance of being affected by strontium^{90} than adults do. Biologists have not had enough time to find out just how harmful certain concentrations of strontium^{90} may be. Strontium^{90} has a half-life of about thirty years. Cesium^{137} is another radioactive waste product. The body "mistakes" cesium^{137} for potassium and concentrates it in muscle tissue. Like strontium^{90}, cesium^{137} has a half-life of about thirty years.

So different kinds of isotopes become concentrated in different parts of the body, and different kinds of organisms tend to store up certain isotopes more than others. Strontium^{90} is taken up mostly by the scales of fish and by the shells of other water animals. Tritium, an isotope of hydrogen, seems to work its way into most parts of most organisms. Certain water plants, algae, can build up concentrations of radioactivity 500,000 times that of the surrounding water; certain insect larvae, 100,000 times; certain fish, 20,000 to 30,000 times. The fact is that our scientists are just now learning how radioactive substances build up in living tissues of different kinds of animals and plants. Also, when we talk about radioactive substances being taken up by different organisms, we must keep in mind that different kinds of organisms have different kinds of diets.

The power companies who are building all the nuclear power plants that produce energy by an atom splitting (fission) process—which is very different from producing energy by nuclear fusion—want us all to believe that nuclear power is "safe" and "clean." These two words appear again and again in their publicity promoting the construction of nuclear power plants. They tell us that the radioactivity they release into the water and air is so dilute, or weak, that it is completely harmless. However, a number of medical scientists feel that they now have enough evidence to show that this is a dangerous lie. The power companies do not agree and continue to claim that nuclear power is clean and safe. Nuclear power is *not* clean. Nuclear power is "safe" only until the first large-scale accident in a nuclear power plant occurs or the first people get cancer from power plant wastes or pollutants.

Where Radioactive Wastes Come from

When our nuclear weapons testing program was started back in the 1940s, a big question was how much damage radioactive wastes would do to living things in the environment. We did not know the answer then, and we still have a lot to learn. Even so, we keep on producing radioactive pollutants at an alarming rate, and we plan to keep right on producing them—through the operation of nuclear power plants—at an alarming rate for many years to come. In addition to the twenty-five or so nuclear power plants now operating in the United States, there are plans to have about sixty more operating before 1980, and maybe five hundred by the year 2000.

The waste pollution story begins in the uranium mines, which are mainly in the West. Colorado, in particular, has a large number of them. Uranium rock from the mines is crushed and processed in mills, where the uranium is separated from the rock. Each ton of uranium ore processed gives up only about five pounds of usable uranium. The remaining 1,995 pounds of waste is radioactive sand called "tailings."

For about twenty-five years now, millions of tons of this radioactive sand have been piling up by river banks or wherever else a uranium mill happens to be located. In the Colorado River Basin alone, more than twelve million tons of tailings have piled up. Blown by the wind and washed into streams feeding the Colorado River, these particular radioactive wastes have polluted the air and water used for drinking and irrigation in parts of California, Nevada, Utah, Wyoming, Colorado, New Mexico, and Arizona.

In at least one Colorado community (Grand Junction) uranium tailings were given away, in ignorance, to builders over a period of fifteen years. The builders used the radioactive sand as foundation fill for schools, private homes, and other buildings. There is now great concern in such communities over how much radiation poisoning people have been getting over the years. Some families have been moved out of their radioactive homes. At one count, eighty buildings in Grand Junction had high radiation levels.

Once uranium is separated from its ore, it is sent to other plants where it is treated in several other ways. Eventually, it ends up as uranium pellets, which are packed into metal rods and become fuel for nuclear reactors, or is processed for use in nuclear bombs.

All along the line, from one processing plant to the next, some radioactive wastes are leaked into the air and other parts of the environment. At any one place or at any one time the amount of radiation leakage may not be very much, but it is accumulating bit by bit, day by day, year by year, century by century . . .

Radiation "Graveyards"

Eventually, the fuel rods for a nuclear reactor go "stale" and have to be replaced. The fuel, in the process of being used, becomes more highly radioactive and is much more dangerous to handle than the fresh uranium pellets originally packed in the rods. The reason is that as the fuel atoms are split their remains turn into radioactive isotopes of nearly twenty-five other elements, most of which give off lethal gamma radiation. Any nuclear power plant has to close down from time to time, gather its radioactive wastes, and pack them up for shipment to a special plant that will reprocess the used fuel and bury or store the other waste materials. Some of the wastes are so highly radioactive that they have to be handled by remote control and placed in lead-lined concrete casks weighing ten tons each. The casks are then trucked by highway or by train to the nearest "graveyard" for radioactive materials, often many hundreds of miles away.

An interesting question is what would happen if there were a train or truck wreck and one of the deadly casks were sent tumbling over the ground. Nuclear power plant engineers will tell you that the casks can withstand nearly any possible accident—for instance, one could be dropped from a height of forty feet and land on one of its corners on a concrete highway without breaking open. There is a "law" in science that says if an accident *can* happen, eventually it *will* happen. *If* one of these deadly casks did break open, within twenty hours people living nearby would be killed by the radioactivity. A much worse situation would be if one of the casks spilled open into a stream, river, or other body of water. The deadly radioactivity would then be carried far and wide. Or think of the possibility of some mentally sick person, like a skyjacker, threatening to blow up a cask when it neared a highly populated area.

When the used fuel is sent to a reprocessing plant, some of the leftover uranium can be separated and used as new fuel, but most of the used fuel is shipped on to a radiation graveyard. The millions of tons of highly radioactive wastes from our bomb factories also are shipped to these same graveyards. Again, radioactive wastes do not simply "go away" like steam. What happens to them once they reach the graveyard?

They collect there and are becoming a greater and greater source of danger to the world. Over the past twenty years or more the United States Atomic Energy Commission has collected somewhere around eighty million gallons of highly dangerous radioactive wastes, not to mention the tons of solid wastes disposed of under its supervision. The liquid wastes alone are enough to poison all of the water on our entire planet! Most of these liquid wastes—so hot from radioactive energy that they boil day and night—are stored in huge concrete and steel tanks in the state of Washington at the Hanford Atomic Producers Operation in Richland. (Each such tank holds from about a half to one and a half million gallons.) Still more is stored in similar tanks in Arco, Idaho, and along

Boiling hot radioactive wastes are stored in million-gallon steel tanks like those being built here by the Atomic Energy Commission at its Hanford Works in southeastern Washington. The U. S. Public Health Service estimates that by 1995 about two billion gallons of radioactive stew will be boiling away in some two hundred such giant tanks.

the Savannah River in South Carolina. According to the United States Public Health Service, by the year 1995 about two billion gallons of AEC licensed radioactive stew will be boiling away in some two hundred tanks. Although the AEC and power companies that own nuclear power plants do not find that thought disturbing, many other people do. One such person is David Lilienthal, who is a former chairman of the Atomic Energy Commission and who had this to say:

These huge quantities of radioactive wastes must somehow be removed from the reactors, must—without mishap —be put into containers that will never rupture; then these vast quantities of poisonous stuff must be moved either to a burial ground or to reprocessing and concentration plants, handled again, and disposed of, by burial or otherwise, with a risk of human error at every step.

If the boiling-hot liquid wastes in the great tanks were not kept stirred by streams of compressed air and cooled by letting the radioactive steam from the hot tanks pass through cooling pipes above the ground, the tanks would blow their tops. They would spray part of their deadly contents into the air and into the rivers.

The exact amounts of radioactive wastes in AEC graveyards are kept secret from the public by the AEC. But it seems likely that each of the 140 or more giant underground storage tanks contains as much radioactivity as has been released by all the nuclear weapons tests since 1945. One frightening question about the graveyards is how *safe* are they? What if a severe earthquake should rock one of the storage areas? What if the cooling system should fail for some reason? These are frightening questions indeed, but the answers could be even more frightening. What if an enemy decided to bomb the storage areas?

As more and more long-lived wastes are collecting in more and more nuclear power plants and radiation graveyards, the AEC is trying to find the best ways of containing and eventually getting rid of them. According to Dr. Joel A. Snow, a physicist at the University of Illinois Center for Advanced Study: "A single gallon of waste would be sufficient to threaten the health of several million people." Yet we read a report saying that one of the AEC's giant tanks leaked "60,000 gallons into the ground before discovery. At least two others are leaking, and leaks are suspected in three or four more." Five years ago the AEC had to ask Congress for $2,500,000 for replacement of failed or failing tanks in its Richland, Washington plant, claiming "there is no assurance that the need for new waste storage tanks can be forestalled."

The Atomic Energy Commission over the years has been working out a way of changing liquid radioactive wastes into solid materials that can be stored in glass-lined steel cylinders buried in abandoned salt mines. Even if this works, though, it

For several years the AEC has been experimenting with its radioactive liquid wastes by cooling them down enough to be changed into solid material. The solids are then stored in glass-lined steel cylinders that are buried in abandoned salt mines in Kansas, never again to be opened by human beings. The cover of one such experimental waste cavity near Hutchinson, Kansas, is shown here.

won't change the amount of radioactive wastes being released into the environment as we build more and more nuclear power plants.

One question about radiation graveyards keeps coming back: HOW SAFE ARE THEY? No one can answer for certain. Perhaps the buried wastes will be safe for a time—for our lifetime. But what about after that? Shouldn't that be our concern, too? The fact is that we do not know nearly as much as we should know about the effects of radioactivity on living things. That is one fact a lot of nuclear engineers overlook.

By the year 2000, if our present speed in building nuclear power plants continues, our grandchildren will be faced with guarding about six billion *curies* of radiation from strontium[90]. That is thirty times as much strontium[90] as would be released in an all-out nuclear war. One curie of strontium[90] is enough to kill a human being. (A "curie" is a unit of radioactivity describing the number of atomic fissions taking place in a particular amount of a particular radioactive element per second.) Tens of billions of curies of radiation from other radioisotopes will also be in storage by the year 2000 and will linger for hundreds and thousands of years. Unlike a rainbow, they will not just go away.

Is There a "Safe" Amount of Radiation?

Man and all other living things have always been exposed to a small amount of radiation. This "background radiation," as it is called, comes from the decay of natural radioactive isotopes in Earth's rocks, soil, and water, and from tiny amounts of these substances that have been taken into our bodies with food, water, and air. Some also comes from cosmic rays reaching us from outer space.

No one knows whether or how much this natural radiation has affected the ways that plants and animals have evolved over the many millions of years of their existence. Our bodies and tissues must have become adapted to this weak radiation, though, or we would not be alive now.

Since 1945, however, we have been adding more and more radioactive wastes to the environment. And this exposes all living things to increasing amounts of radiation.

Some scientists have thought that spreading radioactive wastes thinly through the air and water would weaken their radiation enough so that it would not harm living things. But many scientists—especially biologists—have come to the conclusion that it is very hard to know just what amount of radiation can be considered "safe." Their findings suggest that *any* amount of radiation, however small, can cause *some* damage to living things, even though the damage may not show up for many years. So when we hear someone talking about "safe levels" of radiation, we should ask what they mean by "safe," and "safe" for whom, and "safe" for how long?

INDEX

Air, 12, 14–15, 16
Air pollution, 63, 64, 68, 69, 70, 71
Alchemy (alchemists), 16, 55
Algae, 69
Alpha particles, 42–45; and man-made isotopes, 49–50; and radiation pollution, 63
Aluminum, 49–50; atomic weight of, 50
Animal plankton, 65–68
Animals (animal life), radiation pollution and, 64–76
Anode, 22, 29
Argon^{40}, 51
Artificial (man-made) isotopes, 49–50, 56; half-lives of, 50
Aston, F. W., 47–49
Atomic "ash," 62
Atomic energy, 55–62; from fission, 57–60; from fusion, 60–62; radiation pollution and, 63–76 (see also Radioactive waste materials)
Atomic Energy Commission (AEC), 72–75
Atomic piles (nuclear reactors), 57, 58–60, 61–62; fission, 57–60, 61; fusion, 61–62; and radiation pollution, 63, 65–68
Atomic power, 60. See also Atomic energy; Nuclear power plants
Atomic waste materials, 62, 63–76; radiation pollution and, 63–76 (see also Radioactive waste materials)
Atoms: alpha and beta particles, 42–45 (see also Alpha particles; Beta particles); "complex," 17–19; differing sizes and mass of, 18, 24, 48–49 (see also Mass, atomic; Weight, atomic; specific kinds); early theories of, 11–19 (see also specific scientists); electrons, 23, 24–25, 27–28, 29 (see also Electrons); elements and, 14–19 (see also Elements); energy from, 55–62 (see also Atomic energy); isotopes and, 49–54 (see also Isotopes); kinds of, 11–12; meaning of word, 11; neutrons, 27–28 (see also Neutrons); nucleus of, 25–27 (see also Nucleus, atomic); protons, 27–28 (see also Protons); radioactivity and, 41–46, 45–54 (see also Radioactivity; specific aspects, kinds); recycling (travels in life chain) of, 66–67; self-changing, 41–42; splitting, 55–60; structure of, 45–46; weighing, 48–49

Background (natural) radiation, 76
Bacteria, and recycling of atoms, 67
Barium, 56–57
Becquerel, Henri, 32–37, 39, 40
Beta particles, 42–45; and radiation pollution, 63
Bombs, nuclear, 56–62, 63; fission, 56–60; fusion, 61–62; and radiation pollution, 63–64, 70–76; test ban, 64
Boyle, Robert, 15, 17

Calcite, 18
Calcium (calcium atoms), 18, 36, 66
Calcium carbonate, 67
Cancer, radiation pollution and, 45, 46, 53, 64
Carbon (carbon atoms), 17, 18, 51, 66, 67, 69; isotopes, 49; recycling of, 66–67
Carbon^{12}, 49, 69
Carbon^{14}, 51–53, 69
Carbon dioxide, 15, 67
Cathode, 22, 29
Cathode-ray tube, 29–30, 32
Cell damage, radiation and, 64, 68
Cellulose, 67
Cesium^{137}, 69
Chadwick, James, 44–45
Chain reaction, nuclear, 57, 58–60; fission, 57–60; fusion, 60–62; slowing down of, 60–61
Chicago, University of, world's first atomic pile built at, 59
Clocks, geologic, 51–53
Coal, 66–67
Cobalt^{60}, 46
Colorado, radiation pollution in, 70–71
Colorado River water pollution, 70–71
Columbia River water pollution, 65–68
"Complex atoms," Dalton and, 17, 19
Cosmic rays, 51, 76
"Critical" chain reaction, 60
Crookes, Sir William, 41
Curie, Irène, 49–50
Curie, Jacques, 40
Curie, Marie S., 38, 39–41, 42; death from leukemia of, 41; and discovery of polonium and radium, 39–41, 42; ill., 38
Curie, Pierre, 38, 39, 40–41, 42; ill., 38
Curie(s), radiation, 75–76

Dalton, John, 17–19
Dating, use of half-life of radioactive isotopes for, 51–53
Democritus, 11–15
Deuterium, 60, 61, 62

Earth, the, 14–15, 16; dating, 51–53
Einstein, Albert, 58
Electricity, 21–25, 27, 29, 30, 40, 42–45; from nuclear power, 60
Electrometer, 40
Electrons, 23, 27–28, 29, 55; discovery of, 24–25, 27–28; radiation and, 42–45; radiation pollution and, 63, 64
Elements, 14–19; in the human body, 66, 68–69; known number of, 15, 28; radioactive, 45, 46 (see also specific kinds); radium as "new" element, 40–41
Eskimos, 68

Fallout, radioactive, 64, 68, 69
Feldspar, 51
Fermi, Enrico, 55–58, 59
Fire, 14–15, 16
Fish, radiation pollution and, 65–68, 69

ROY A. GALLANT is well known for his many science books for young people. Author of the best-selling EXPLORING series, among which is *Exploring the Universe* (which won a Thomas Alva Edison Foundation Award), Mr. Gallant is a member of the faculty of The American Museum-Hayden Planetarium and has taught astronomy at the Hackley School (Tarrytown, New York). His biography of Charles Darwin won an NSTA nomination as one of the best science books published for young people in 1972. Mr. Gallant is also a principal author of the Ginn Science Program, a major elementary science textbook program. In addition to his writing and teaching, Mr. Gallant serves as a science education consultant to a number of publishers. He lives with his family in their mountain retreat in Rangeley, Maine. He is a graduate of Bowdoin College and Columbia University, where he did postgraduate work and also was a member of the faculty.